徐嘉儀 編著

新手入廚系列

簡易粉麵

前言

縱使生活繁忙，下班後也不一定要吃即食麵。

粉麵是日常飲食其中一種最方便的食材，源自穀類，是熱量的主要來源；別看烹調粉麵不過簡單的步驟，其中每個細節都會影響其賣相和口感，要是忽略了就煮不出最佳的風味。

本書精選三十多道家常麵食，既有道地口味，也有異國風味，包括蛋麵、上海麵、公仔麵、通心粉、意粉、米粉、米線、河粉、烏冬、伊麵、拉麵、年糕和粉皮。做法簡單，食材易找，還附有材料處理方法，一書在手，不論是早餐、午餐、晚餐、下午茶、宵夜，都能大飽口福，天天享受家常麵食。

目錄

蛋麵 · 上海麵

公仔麵 · 通心粉 · 意粉

炒蛋麵

色澤淺黃，含鹼味，乾爽，沒有黑點或發霉，適合炒和半煎炸。

Stir-fried egg noodles

Light yellow, alkaline flavor, dry, no stain / black spot or mould, suitable for stir-frying and semi-frying.

幼蛋麵

色澤呈深黃，麵條纖幼，柔軟乾爽，沒有黑點或發霉現象，適合焯煮。

Thin egg noodles

Dark yellow, thin, soft and dry, no stain / black sport or mould, suitable for boiling.

上海粗麵

白麵條，粗大，外層有薄粉，
屬半濕麵，適合炒和燜。

Shanghai coarse noodles
White noodles, coarse,
outer part / surface with
some flour, semi-wet,
suitable for stir-frying and
stewing.

上海幼麵

屬半濕白麵條，質地幼細，
麵條纖幼，適合放湯、拌麵
和焯煮。

Shanghai thin noodles
Semi-wet white noodles,
delicate texture, thin,
suitable for serving in soup,
stir with sauce and boiling.

上海中麵

屬半濕白麵條，外有薄粉，質
地柔軟，適合焯、煮和放湯。

Shanghai middle noodles
Semi-wet white noodles,
outer part / surface with
some flour, soft, suitable for
boiling, cooking and serving
in soup.

米粉

米與清水磨成米漿，乾燥而成，幼身乾爽，含米香，適合炒和燜。

Rice vermicelli

Grind rice and water to produce rice paste, let dry, thin and dry with rice flavor, suitable for stir-frying and stewing.

米線

潔白柔軟，米味濃郁，有彈力，適合放湯和焯煮。

Rice vermicelli

White, soft, rich rice flavor, elastic, suitable for serving in soup and boiling.

炒河粉

粉片纖薄，粉條闊約 3/4 厘米，沒有異味，表面有油份，適合炒。

Stir-fried rice noodles

Thin, its width is around 3/4 cm, no odour / bad smell, surface contains oil, suitable for stir-frying.

烏冬

粉質軟滑，麵條呈方形，潔白富彈力，適合炒、焯、煮和燜。

Udon

Rich in starch, soft, square shape, white, elastic, suitable for stir-frying, boiling, cooking and stewing.

伊麵
油炸麵條，質地膨鬆，呈淡黃色，沒有「油膩」味道。

E-fu noodles
Deep-dried noodles, expanding and loose texture, light yellow, no bad "oily" flavor.

拉麵
麵質有點像中式的油麵，不肥膩，質地富彈力，含鹼味，適合焯煮和炒。

Ramen
Similar to the texture of Chinese oily noodles, not greasy, elastic, salty, suitable for boiling and stir-frying.

年糕
潔白軟滑，質地細緻，有嚼勁，適合放湯、炒和燜。

Rice cake
White, soft, thin / delicate, chewing, suitable for serving in soup, stir-frying and stewing.

洋葱
外衣金黃呈光澤，結實飽滿，
完整無傷。

Onion
Golden brown and shiny,
firm and rich, no wound.

金菇
菇帽細小，沒有瘀傷，堅挺含
水份。

Enoki mushroom
Small mushroom top / cap,
no wound, firm and with
moisture.

韭黃
顏色青白帶微黃，飽滿含
水份，沒有腐爛。

Yellow chive
White and light yellow,
rich and with moisture,
not rotten.

銀芽
去掉豆芽和尾部的銀白豆梗，
不瘀黑或糜爛，堅挺新鮮。

Sliver sprout
Remove bud and end part,
firm and fresh without black
or rotten part.

中國芹菜
味道濃郁，色澤翠綠，堅挺沒有瘀傷。

Chinese celery
Rich flavor, green body, firm and without wound.

中國蒜
頭部潔白，綠色部份翠綠，硬實堅挺。

Chinese garlic
White head, green, firm.

獨子蒜
潔白硬實，沒有蟲口或腐爛，乾爽完整。

Single clove garlic
White and firm, no damage caused by pest, not rotten, complete and dry.

紅頭蔥
頭部紫紅，氣根嫩脆，頭部渾圓飽滿，全身翠綠堅挺。

Red onion
Purple red head, delicate root, rich and round head, green body and firm.

1. 乾麵類如蛋麵、生麵和麵餅等，看似乾爽，但貯藏期很短，宜放在密封瓶或保鮮袋內，置陰涼地方擺放，可貯 2 至 3 天。

2. 包裝的烏冬、拉麵、米粉可放陰涼地方置放，但開封後，就要放在密實盒或保鮮袋。

3. 新鮮麵類如米線、河粉和伊麵就要即買即用，而首兩款就要置冰箱候用，用時才取出，否則很容易變壞。

4. 上海麵類未用時，以紙包裹，置放在保鮮盒內可待 1 至 2 天，如放進冷凍格可儲存 1 星期；但冰過的麵條容易折斷，麵質會比較脆軟。

1. Dried noodles such as whole egg noodles, raw noodles and noodle cakes are dry but the storage time is short. They can be kept for 2-3 days if store in airtight container or plastic bag and cool, dim-lighted place.

2. Packed Udon, ramen, rice vermicelli should be kept in cool, dim-lighted place. They should be kept in airtight container or plastic bag after opening.

3. Fresh noodles such as rice vermicelli, rice noodles and E-fu noodles should be used once purchased. Rice vermicelli and rice noodles should be kept in fridge before use. Otherwise, they will deteriorate easily.

4. Shanghai noodles can be kept for 1-2 days if wrap in paper and stored in airtight container before use. They can be kept for 1 week if stored in fridge. However, frozen noodles are easily broken and the texture is pretty sticky and soft.

製作魚湯 Cooking fish soup

煮麵缺不了魚湯，營養豐富又美味，在市場購買一些新鮮冰鮮小魚，湯頭甜美，價錢便宜又鮮甜，可預先準備，置冰箱備用。

Fish soup is indispensible to noodle cooking as it is rich in nutrients and delicious. You may buy small fresh fishes in market to cook the soup. The cost incurred is low and the soup is sweet. You may prepare the soup in advance and keep in fridge.

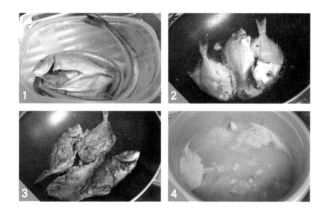

做法 | Method

1. 海魚劏洗乾淨，用少許鹽抹擦全身。
2. 放入已預熱的鑊中，下油燒熱，放入已劏洗的魚。
3. 把魚煎至金黃。
4. 放入滾水中熬煮至奶白色。

1. Remove intestine / unnecessary parts of sea fishes and wash. Rub with little salt.
2. Put in preheated wok with oil.
3. Pan fry until golden brown.
4. Put into hot water and boil until the soup turns milky.

焯意大利粉 Cooking spaghetti

意大利粉要煮得柔軟適中，兼有嚼口，就要懂得如何處理。包裝上的說明，只給了烹煮時間，但有些細節如看粉條的質感和過冷河，才是成功的關鍵。

Cooking spaghetti until its degree of softness is proper. The instruction shown on packing provides cooking time only. It is important to observe the texture of spaghetti and pay attention to the process of blanching.

◯◯ 做法 ｜ Method

1. 煲水一鍋，水必須能蓋過面為準則，還要處於沸騰狀態。

2. 放入意大利粉，如果麵條不能完全放鍋中，待其軟身，再用手輕輕按入煲內。

3. 揭鍋蓋煮意大利粉至熟，中間完全透明，需時約 8~10 分鐘，按麵條粗幼調節烹調時間。

4. 當麵條煮熟，其質感會略發脹兼軟身，麵水會呈奶白色，取一條麵條弄斷，中間完全透明不實心。

5. 倒去麵水，放水喉下沖水至清爽，瀝水便可使用。

1. Boil a pot of water. The water must be able to cover the surface and in boiling state.

2. Put in spaghetti and boil until soft. Slightly press spaghetti into water if it cannot be put into the pot and.

3. Remove cover and cook spaghetti for 8-10 minutes until well done (the middle part should be transparent). The cooking time depends on the thickness of spaghetti.

4. When spaghetti is well done, it slightly expands and is soft. At the same time, boiling water of spaghetti should be milky. Get a string of spaghetti and break it. The middle part should be transparent.

5. Drain away boiling water of spaghetti. Rinse spaghetti under tap water to make it crispy and fresh. Drain before use.

半煎麵 Semi-cooked pan fried noodles

肉絲炒麵用淡黃的麵餅，其質地柔軟而麵條纖細，需要用半煎炸烹調法處理，避免麵條很油膩，可用易潔鑊處理，省時省油省燃料，但效果不錯。

Use light yellow noodle cakes to make stir-fried noodles with shredded meat. As the texture of noodle cakes are soft and delicate, it is necessary to semi deep-fry or pan fry. You may use non-stick wok or pan to avoid greasy texture as well as to save time and oil.

◯◯◯ 做法 │ Method

1. 預先燒一鍋清水。
2. 放入麵餅，煮開後用筷子挑散麵條，倒出麵條，放入筲箕內瀝乾，過冷河。
3. 熱鑊下 2~3 湯匙油，放入已過冷的麵條，煎至乾脆，翻轉，續煎至兩面金黃香脆。
4. 上碟後瀝油，便可使用。

1. Boil water.
2. Put in noodle cakes. Use chopstick to separate strings of noodles. Sieve and drain noodles, then blanch.
3. Heat wok with 2-3 tbsps of oil. Add in blanched noodles. Pan fry until crispy. Turn over, pan fry until both sides are golden brown.
4. Pour some oil on top when serve.

4~6 人
Serves 4~6

20 分鐘
20 minutes

肉絲炒麵
Stir-fried Noodles
with Shredded Meat

⬤⬤⬤ 材料 | Ingredients

蛋麵 2 個	2 pcs egg noodles
冬菇 4 朵（切絲）	4 pcs dried black mushrooms (shredded)
豬肉絲 80 克	80g shredded pork
銀芽 80 克	80g silver sprouts
韭黃 20 克	20g yellow chive
蒜茸 1 茶匙	1 tsp minced garlic

入廚貼士 | Cooking Tips

- 用易潔鑊煎蛋麵，容易處理兼外形漂亮。
- It is easier and more convenient to pan fry egg noodles by using non-sticky wok.

⊙⊙⊙ 醃料 | Marinade

鹽 1/2 茶匙
糖 1/2 茶匙
生抽 1/2 茶匙
生粉 1/2 茶匙

1/2 tsp salt
1/2 tsp sugar
1/2 tsp light soy sauce
1/2 tsp cornstarch

⊙⊙⊙ 調味料 | Seasonings

上湯 180 毫升
蠔油 2 茶匙
鹽 1/2 茶匙
糖 1/4 茶匙
胡椒粉少許

180 ml broth
2 tsps oyster sauce
1/2 tsp salt
1/4 tsp sugar
Pinch of pepper

⊙⊙⊙ 做法 | Method

1. 燒滾水，加入蛋麵及 1 湯匙油將麵煮散，撈起待用。

2. 把醃料放在肉絲內撈勻，醃 15 分鐘，稍焯後待用。

3. 冬菇絲加 3 湯匙上湯蒸 15 分鐘；銀芽稍焯；韭黃切成約 1 吋的段，備用。

4. 燒熱鑊，加 2 湯匙油，用慢火把麵煎至兩面金黃，上碟保暖。

5. 燒熱鑊，爆香蒜茸，加入豬肉絲、銀芽和冬菇絲，炒勻後潷酒，加調味料，再下生粉水，稍微煮稠，最後下韭黃，淋於麵上。

1. Boil water. Add in egg noodles and 1 tbsp of oil. Cook until well done and noodles are loose. Drain and set aside.

2. Marinate shredded pork for 15 minutes and scald for a while.

3. Steam dried black mushrooms with 3 tbsps of broth for 15 minutes. Boil silver sprouts for a while. Cut yellow chive into 1 inch long. Set aside.

4. Heat wok with 2 tbsps of oil. Pan fry egg noodles over low heat until both sides are golden brown. Put onto a plate to keep warm.

5. Heat wok. Saute minced garlic, add in shredded pork, silver sprouts and shredded dried black mushrooms, stir-fry well. Sprinkle wine, stir in seasonings, cornstarch solution and cook until slightly thickened. Add in yellow chive. Pour into a bowl with noodles.

豉油皇炒麵

Stir-fried Cantonese Noodles with Soy Sauce

材料 | Ingredients

炒麵餅 2 個
銀芽 150 克
火腿 3 片
韭黃 2 條
麻油 1 湯匙

2 pcs cake noodle
150g silver sprouts
3 slices ham
2 sprigs yellow chive
1 tbsp sesame oil

2~4 人
Serves 2~4

15 分鐘
15 minutes

燴汁 | Sauce

生抽 1 1/2 湯匙
老抽 1 湯匙
糖 1 茶匙
清水 2 湯匙

1 1/2 tbsps light soy sauce
1 tbsp dark soy sauce
1 tsp sugar
2 tbsp water

入廚貼士 | Cooking Tips

- 把麵和麻油一同撈勻，炒麵時容易散開，不糊在一起。
- The noodles will not easily stick together during stir-frying if mixed with sesame oil in advance.

蛋麵．上海麵

Egg Noodles & Shanghai Noodles

做法 | Method

1. 韭黃及火腿分別切段。
2. 煲滾水，放下麵餅，用筷子弄散，馬上取出過冷河，瀝乾，加 1/2 湯匙油，拌勻。
3. 熱鑊燒熱油，爆香銀芽、韭黃段和火腿段炒熟，盛起。
4. 原鑊下麵，加入燴汁炒至略為收乾，加入銀芽、韭黃和火腿炒勻，熄火，下麻油拌勻即成。

1. Chop yellow chive and ham into sections.
2. Boil cake noodles in hot water. Separate strings of noodles by chopsticks. Take out, blanch and drain. Mix with 1/2 tbsp of oil.
3. Heat wok with oil and sauté silver sprouts, chopped yellow chive and ham. Stir-fry until well done. Set aside.
4. Heat wok again, add in noodles and sauce. Stir-fry until the sauce is slightly thickened. Add in silver sprouts, yellow chive and ham. Remove from heat. Mix well with sesame oil and serve.

蜜味豬頸肉湯麵

Noodles in Soup with Honey Pork Jowl

材料 | Ingredients

全蛋麵 150 克
豬頸肉 2 片（約 450 克）
麥芽糖 3 湯匙
上湯適量

150g whole egg noodles
2 pcs (about 450g) pork jowl
3 tbsps maltose
Some broth

22

⊙⊙ 醃料 | Marinade

叉燒醬 3 湯匙
蒜茸 2 茶匙
生抽 2 茶匙
乾葱茸 1 茶匙
麻油少許

3 tbsps roast pork sauce
2 tsps garlic
2 tsps light soy sauce
1 tsp minced shallot
 Some sesame oil

入廚貼士 | Cooking Tips

- 讓豬頸肉快點熟,可用刀在厚肉部份剟紋,容易入味兼省時烹煮。
- Slightly slice thick pork jowl to save cooking time and make it absorbing sauce easily.

⊙⊙ 做法 | Method

1. 醃料放豬頸肉內撈勻,醃 5 小時。
2. 預熱焗爐,以 180℃ 將豬頸肉烤焗 15 分鐘,掃上麥芽糖,再焗約 5 分鐘,切薄片。
3. 蛋麵放熱水中煮熟,撈出,放入深碗中。
4. 煮滾上湯,倒入深碗中,放上豬頸肉片,即可食用。

1. Marinate pork jowl for 5 hours.
2. Bake pork jowl in a preheated oven at 180℃ for 15 minutes. Brush maltose, bake for another 5 minutes and slice.
3. Boil egg noodles in hot water, drain and put into a bowl.
4. Boil broth and pour into a bowl, put sliced pork jowl on top and serve.

咖喱魚蛋豬皮車仔麵

Stall Noodles with Curry Fish Ball and Pig Skin

材料 | Ingredients

粗麵 4 個	4pcs coarse noodles
白蘿蔔 450 克	450g radish
已浸發豬皮 225 克	225g soaked pig skin
炸魚蛋 150 克	150g deep-fried fish balls
咖喱醬 2 湯匙	2 tbsps curry paste
磨豉醬 1 湯匙	1 tbsp crushed yellow bean sauce
蒜茸 1 茶匙	1 tsp minced garlic
薑茸 1 茶匙	1 tsp ginger puree
上湯 4 杯	4 cups broth

4 人
Serves 4

25 分鐘
25 minutes

⊙⊙ 調味料 | Seasonings

生抽 1 湯匙
糖 1 茶匙
鹽 1/2 茶匙
胡椒粉少許
麻油少許
清水 2 1/2 杯

1 tbsp light soy sauce
1 tsp sugar
1/2 tsp salt
Pinch of pepper
Some sesame oil
2 1/2 cups water

⊙⊙ 做法 | Method

1. 將粗麵放滾水煮 2 分鐘，撈起，放入深碗內。
2. 白蘿蔔去皮切塊，放滾水內煮 3 分鐘，盛起待用。
3. 豬皮切細塊，放滾水內煮 1 分鐘，盛起待用。
4. 燒熱鑊，加 2 湯匙油，爆香蒜茸及薑茸，加入磨豉醬及咖喱醬，略炒香。
5. 注入調味料，煮滾後加入魚蛋、白蘿蔔和豬皮，煮 15 分鐘熄火，加蓋焗 1 小時。
6. 煮滾上湯，倒入深碗，加入咖喱魚蛋、白蘿蔔和豬皮即可供食。

1. Boil coarse noodles in hot water for 2 minutes. Drain and pour into a bowl.
2. Peel radish and cut into pieces. Boil in hot water for 3 minutes, drain and set aside.
3. Chop pig skin into small pieces. Boil in hot water for 1 minute, drain and set aside.
4. Heat wok with 2 tbsps of oil. Sauté minced garlic and ginger puree, add in crushed yellow bean sauce and curry paste. Stir-fry for a while.
5. Stir in seasoning and cook until well done. Add in fish balls, radish and pig skin, cook for 15 minutes. Turn heat off and cover for 1 hour.
6. Boil broth, pour into a bowl, add in curry fish balls, radish and pig skin. Serve.

入廚貼士 | Cooking Tips

- 浸發豬皮在清洗時，加點鹽浸 10 至 15 分鐘，再沖洗，然後焯煮，可去除異味。
- Wash pig skin and soak in water with some salt for 10-15 minutes. Wash and blanch to get rid of bad smell / odor.

時菜豬膶湯麵

Noodles in Soup with Vegetables and Pork Liver

◯◯◯ 材料 | Ingredients

上海幼麵 2 個
時菜 450 克
豬膶片 150 克
蒜頭 4 粒
薑 1 片
上湯 750 毫升

2 pcs Shanghai thin noodles
450g vegetables
150g pig's liver
4 cloves garlic
1 slice ginger
750 ml broth

⟨⟨⟨ 芡汁 | Thickening

生粉 2 茶匙	2 tsps cornstarch
生抽 1 茶匙	1 tsp light soy sauce
紹酒 1 茶匙	1 tsp Shaoxing wine
薑汁 1 茶匙	1 tsp ginger juice
糖 1/3 茶匙	1/3 tsp sugar

⟨⟨⟨ 做法 | Method

1. 把上海幼麵放熱水內煮 5 分鐘，撈起，放入深碗。
2. 將約 500 毫清水煮滾，放入豬膶片，熄火焗 1 分鐘。
3. 取出豬膶片，放入冰水略浸。
4. 燒熱鑊，下油，爆香薑片及蒜頭，加入時菜炒熟，注入上湯，煮滾後倒入碗裏即可。

1. Boil Shanghai thin noodles in hot water for 5 minutes, drain and put into a bowl.
2. Boil 500 ml of water and add in pig's liver. Turn heat off and cook for 1 minute with the lid covered.
3. Take pig's liver out and soak in ice water for a while.
4. Heat wok with oil. Sauté sliced ginger and garlic. Add vegetables and stir-fry until well done. Add in broth and cook until well done. Pour into a bowl.

入廚貼士 | Cooking Tips

- 豬膶含血水，加入薑汁和酒可辟去腥味，但離水久了會再次滲出血水。
- Pig's liver contains blood water. Add ginger sauce and wine to get rid of bad smell / odor. However, blood water will flow out again if pig's liver is taken away from water for a long period of time.

擔擔麵

Sichuan Dandan Noodles

材料 | Ingredients

上海幼麵 300 克	300g Shanghai thin noodles
免治豬肉 75 克	75g minced pork
蝦米碎 2 湯匙	2 tbsps chopped dried shrimps
豆瓣醬 2 湯匙	2 tbsps broad bean paste
乾葱茸 2 湯匙	2 tbsps shallot puree
榨菜粒 2 湯匙	2 tbsps diced preserved vegetables
葱粒 2 湯匙	2 tbsps diced spring onion
蒜茸 1/2 茶匙	1/2 tsp minced garlic
薑茸 1/2 茶匙	1/2 tsp ginger puree
上湯 1 公升	1 litre broth

醃料 | Marinade

生抽 2 茶匙	2 tsps light soy sauce
糖 1 茶匙	1 tsp sugar
鹽 1 茶匙	1 tsp salt

2~4 人
Serves 2~4

15 分鐘
15 minutes

⊙ 調味料 | Seasonings

麻醬 1 湯匙	1 tbsp sesame paste
麻油 1 茶匙	1 tsp sesame oil
生抽 1 茶匙	1 tsp light soy sauce
糖 1/2 茶匙	1/2 tsp sugar
清水 2 湯匙	2 tbsps water

⊙ 做法 | Method

1. 把上海幼麵放滾水煮 5 分鐘，撈起，放入深碗裏。
2. 免治豬肉加入醃料拌勻，醃 1/2 小時左右。
3. 在鑊中下少許油，先爆香乾葱茸、蒜茸、薑茸、蝦米碎和榨菜粒，隨後加入免治豬肉，煮熟。
4. 加入調味料煮滾，再撒入少許葱花炒勻，撈起。
5. 注入上湯煮滾，分別倒入碗中，將已炒好的材料放在上面，即可食用。

1. Boil Shanghai thin noodles in water for 5 minutes. Drain and pour into a bowl.
2. Marinate minced pork for 30 minutes.
3. Heat wok with some oil. Sauté shallot puree, minced garlic, ginger puree, chopped dried shrimps and spicy diced preserved vegetables. Add in minced pork and cook until well done.
4. Stir in seasoning and boil. Add in some spring onion, stir-fry and set aside.
5. Add broth and bring to a boil. Pour into a bowl, place stir-fried ingredients on top. Serve.

入廚貼士 | Cooking Tips

- 免治豬肉不能清洗，否則會令豬肉含水份太多。最好買半肥瘦豬肉回來自己用刀剁，比較衞生。
- Do not wash minced pork, otherwise it will contain too much water. It is better to buy semi-fat pork and chop it on your own for better hygiene.

榨菜肉絲湯麵

Noodle in Soup with Preserved Vegetables and Shredded Pork

ⓒⓒⓒ 材料 | Ingredients

上海麵 300 克
豬肉絲 200 克
榨菜絲 50 克
紅椒絲 2 隻
蒜茸 1 茶匙
上湯 1 1/2 公升

300g Shanghai noodles
200g shredded pork
50g shredded preserved vegetables
2 pcs shredded red chili
1 tsp minced garlic
1 1/2 litres broth

醃料 | Marinade

生抽 2 茶匙
生粉 2 茶匙
胡椒粉少許
麻油少許

2 tsps light soy sauce
2 tsps cornstarch
Pinch of pepper
Some sesame oil

入廚貼士 | Cooking Tips

* 上海麵屬白麵條，烹煮時會令清水變滑潺潺，過冷河會清爽一點，但在放湯前需放熱水稍焯。
* Shanghai noodle is a type of white noodle. The water of boiling noodles will become smooth and milky. Blanch noodles to make them fresh and crispy. Boil it hot water for a while before pouring into the soup.

做法 | Method

1. 上海麵放熱水中煮熟，撈起，放入深碗中。
2. 洗淨榨菜絲，略浸。醃料加入豬肉絲內撈勻，醃 30 分鐘。
3. 燒熱油，放紅椒絲爆香，加入豬肉絲炒熟，再加入榨菜絲及 1/2 茶匙糖兜炒，備用。
4. 煮滾上湯，放入上海麵，煮熟後倒在碗中，放上榨菜肉絲，即可食用。

1. Boil Shanghai noodles in hot water until well done. Drain and pour into a bowl.
2. Wash shredded preserved vegetables and soak for a while. Marinate shredded preserved pork for 30 minutes.
3. Heat wok with oil. Sauté shredded red chili, add shredded pork and stir-fry, then shredded preserved vegetables and 1/2 tsp of sugar. Stir-fry and set aside.
4. Boil broth, add Shanghai noodles and cook until well done. Pour into a bowl, place shredded preserved vegetables on top and serve.

嫩雞青菜燴麵

Stewed Noodles with Chicken and Vegetables

◯◯◯ 材料 | Ingredients

上海幼麵 2 個（煮熟過冷河）	2 pcs Shanghai thin noodles (boil and blanch)
小棠菜 300 克	300g Shanghai white cabbage
雞肉 150 克	150g chicken
濃雞湯 5 杯	5 cups condensed chicken soup
鹽 1/2 茶匙	1/2 tsp salt

◯◯◯ 醃料 | Marinade

鹽 1/2 茶匙	1/2 tsp salt
胡椒粉少許	Pinch of pepper
麻油少許	Some sesame oil
雞粉少許	Some chicken powder

4~6 人
Serves 4~6

30 分鐘
30 minutes

⦾ 濃雞湯 | Condensed chicken soup

冰鮮雞 1 隻（約 1350 克）	1 pc (1350g) frozen chicken
排骨 600 克	600g spare rib
金華火腿 75 克	75g Jinhua ham
清水 6 公升	6 litres water

⦾ 做法 | Method

1. 準備雞湯：冰鮮雞洗淨，去清內臟，用少許鹽擦勻全身待片刻，汆水過冷。燒水一鍋，然後把濃雞湯材料放入煲內以大火滾起，轉中火煲 2 1/2 小時，調高火力，大滾 30 分鐘，涼凍後撇去浮面雜質和雞油便可。
2. 雞肉洗淨，切粒，加入醃料撈勻，醃 10 分鐘。
3. 把小棠菜洗淨、切粒。
4. 煮滾濃雞湯，加入雞肉粒煮熟，放上海麵煮至稍軟時，加入小棠菜拌勻，下鹽調味。

1. Prepare chicken soup: Wash frozen chicken, remove organs, rub with some salt for a while and blanch. Boil a pot of water, add ingredients of condensed chicken soup and cook over high heat. Adjust to medium heat and boil for 2 1/2 hours. Turn to high heat and boil for 30 minutes. Leave to cool, remove impurities and chicken oil on top.
2. Wash chicken, dice and marinate for 10 minutes.
3. Wash Shanghai white cabbage and dice.
4. Boil condensed chicken soup, add diced chicken and cook until well done. Add Shanghai noodles and cook until pretty soft. Add Shanghai white cabbage and season with salt.

入廚貼士 | Cooking Tips

- 雞湯清澈，放涼後撇去浮面雜質，不要把杓子胡亂攪動，令沉澱物浮起，只取面層的湯用。
- Chicken soup should be clear. Remove impurities on surface after cool. Do not move / stir wooden ladle improperly to make sediments on the bottom floating to surface. Use top layer of soup only.

雜菌魚湯燴麵

Stewed Noodles with Mixed Mushrooms in Fish Soup

◯◯◯ 材料 | Ingredients

上海幼麵 2 個（煮熟過冷河）
雞髀菇 50 克（切片）
秀珍菇 50 克
鮮冬菇 50 克（切片）
鮮蘑菇 50 克（切片）
濃魚湯 5 杯

2 pcs Shanghai thin noodles (boil and blanch)
50g chicken thigh mushrooms (sliced)
50g oyster mushrooms
50g fresh black mushrooms (sliced)
50g fresh button mushrooms (sliced)
5 cups condensed fish soup

◯◯◯ 濃魚湯 | Condensed fish soup

冰鮮魚 900 克	900g frozen fish
薑 1-2 片	1-2 pcs ginger (sliced)
葱 1-2 條	1-2 sprigs spring onion
清水 3 公升	3 litres water

◯◯◯ 做法 | Method

1. 準備濃魚湯：把魚洗淨，用少許鹽擦勻，待片刻。熱鑊燒，加 2-3 湯匙油，放下魚煎至兩面金黃色，放入魚袋，加入清水煮至濃縮約 5 杯，需時約 1 小時。

2. 燒滾水，放入雞髀菇、秀珍菇、鮮冬菇和鮮蘑菇，焯一會，過冷河備用。

3. 注入 5 杯濃魚湯，煮滾後加入上海幼麵，稍微煮至軟。

4. 加入菇類拌勻，下鹽調味，即可盛碗供食。

1. Prepare condensed fish soup: Wash fish, rub with some salt and leave for a while. Heat wok with 2-3 tbsps of oil, pan fry fish until both sides are golden brown. Put into a fish bag, add water and boil for 1 hour until 5 cups of water left.

2. Boil water. Add chicken thigh mushrooms, oyster mushrooms, fresh black mushrooms and fresh button mushrooms. Boil for a while and blanch.

3. Add 5 cups of condensed fish soup and bring to a boil, add Shanghai noodles and cook until soft.

4. Add mushrooms and mix well, stir in seasonings and serve.

入廚貼士 | Cooking Tips

- 煮白麵條時不要需要蓋蓋，然後過冷河，可保持麵質清爽有韌度。
- Cook white noodles without covering the lid and blanch to keep texture fresh and elastic.

乾炒豬扒公仔麵

Stir-fried Instant Noodles with Pork Chop

材料 | Ingredients

即食麵 2 個	2 packs instant noodles
豬扒 2 塊	2 slices pork chop
洋葱 1/2 個（切絲）	1/2 onion (shredded)
葱 2 條（切絲）	2 sprigs spring onion (shredded)
銀芽 50 克	50g silver sprouts
蒜茸 1 茶匙	1 tsp minced garlic

醃料 | Marinade

生抽 2 茶匙	2 tsps light soy sauce
糖 1 茶匙	1 tsp sugar
鹽 1 茶匙	1 tsp salt

調味料 | Seasonings

即食麵麻油 2 包	2 packs instant noodle sesame oil
即食麵湯包 1 包	1 pack instant noodle sauce
老抽 2 茶匙	2 tsps dark soy sauce

葱油 | Spring onion oil

葱 1 條	1 sprig spring onion
油 3 湯匙	3 tbsps oil
雞粉 1/4 茶匙	1/4 tsp chicken powder

做法 | Method

1. 把葱切粒，拌入雞粉，灒滾油，待用。
2. 燒滾水，放入即食麵煮 2 分鐘，撈起，過冷河，瀝乾，加入調味料，拌勻待用。
3. 用刀背剁鬆豬扒，放入醃料內撈勻，醃 2 小時。
4. 燒熱鑊，放入豬扒煎熟，盛起，切條，備用。
5. 熱鑊下油，爆香洋葱絲和蒜茸，加入銀芽及即食麵炒勻，最後加入豬扒條及葱絲，快速炒勻上碟，灑上葱油即成。

1. Dice spring onion, stir in chicken powder and add in hot oil.
2. Boil instant noodles in hot water for 2 minutes. Take out, blanch, drain, add in seasonings, and mix well.
3. Use back of knife to pound pork chop. Marinate for 2 hours.
4. Heat wok. Pan fry pork chop until well done. Chop into pieces and set aside.
5. Heat wok with oil. Sauté shredded onion and minced garlic, add in silver sprouts and instant noodles and stir-fry for a while. Add chopped pork chop and spring onion at last. Stir-fry quickly and put onto a plate. Sprinkle spring onion oil on top.

入廚貼士 | Cooking Tips

- 即食麵要煮腍一點，因為經熱炒後會略硬些。
- Boil instant noodles until soft as it will be pretty firm after stir-frying.

材料 | Ingredients

公仔麵 4 個	4 packs instant noodles
免治豬肉 450 克	450g minced pork
洋葱 1 個（切絲）	1 onion (shredded)
紅椒 3 隻（切絲）	3 red chili (shredded)
乾葱茸 4 茶匙	4 tsps minced shallot
油咖喱 3 湯匙	3 tbsps oily curry paste
上湯 625 毫升	625 ml broth
椰汁 1/2 杯	1/2 cup coconut milk
花奶 1/4 杯	1/4 cup evaporated milk
鹽 1/8 茶匙	1/8 tsp salt

4 人
Serves 4

20 分鐘
20 minutes

◎◎◎ 醃料 | Marinade

生粉 2 湯匙	2 tbsps cornstarch
鹽 1/2 茶匙	1/2 tsp salt
糖 1/2 茶匙	1/2 tsp sugar
生抽 1/4 茶匙	1/4 tsp light soy sauce
雞蛋 1/2 隻（後下）	1/2 pc egg (add lastly)
胡椒粉少許	Pinch of pepper
清水 3 湯匙（分次後下）	3 tbsps water (add lastly and by batches)

◎◎◎ 做法 | Method

1. 除清水外，把醃料放進免治豬肉裏，一邊攪拌；一邊慢慢放入 3 湯匙水，醃約 15 分鐘。

2. 把免治豬肉做成約直徑 1 吋的肉丸，燒熱油鍋，放入豬肉丸炸至金黃，備用。

3. 燒熱鍋，下油，爆香洋葱絲及乾葱茸，再加紅椒絲和油咖喱，略炒片刻至有香味，倒入上湯及肉丸，煮約 10 分鐘至汁液濃稠。

4. 燒熱一鍋水，放入公仔麵煮熟，盛起，瀝乾，將咖喱肉丸放上面即成。

1. Add all marinade (except water) into minced pork, mix well; add in 3 tbsps of water slowly and leave for 15 minutes.

2. Shape minced pork into balls of 1 inch in diameter. Deep-fry in hot oil till golden brown. Set aside.

3. Heat wok with oil. Sauté shredded onion and minced shallot, add red chili and oily curry paste, stir-fry for a while till aroma comes out. Add broth and meat balls and cook for 10 minutes until sauce is thickened.

4. Boil instant noodles in hot water, drain and pour curry meat balls on top.

入廚貼士 | Cooking Tips

- 肉丸經油炸後容易定型兼已熟，與醬汁同煮後快就可上桌。
- After deep-frying, meat balls are firm and well done. Cook with sauce and serve.

南瓜肉碎通心粉

Macaroni with Pumpkin and Minced Pork

材料 | Ingredients

通心粉 100 克
南瓜粒 200 克
豬肉碎 150 克
上湯 1 公升

100g macaroni
200g diced pumpkin
150g minced pork
1 litre broth

醃料 | Marinade

生抽 1 1/2 茶匙	1 1/2 tsps light soy sauce
生粉 1 茶匙	1 tsp cornstarch
紹酒 1 茶匙	1 tsp Shaoxing wine
糖 1/4 茶匙	1/4 tsp sugar
麻油及胡椒粉各少許	Some sesame oil and pepper
清水 2 茶匙	2 tsps water

做法 | Method

1. 通心粉放熱水煮約 5~8 分鐘至熟，撈起，放入深湯碗中。
2. 在滾水裏放入少許鹽，再放南瓜粒，煮軟，撈起備用。
3. 燒熱鑊，加油，炒熟肉碎，加入南瓜粒炒勻。
4. 煮滾上湯，倒入盛有通心粉的湯碗中，加入已炒熟的南瓜肉碎，即可食用。

1. Boil macaroni in hot water for 5-8 minutes. Drain and pour into a deep bowl.
2. Put some salt in hot water, add diced pumpkin and cook until soft, drain and set aside.
3. Heat wok with oil, stir-fry minced pork. Then add diced pumpkin and stir well until well done.
4. Boil broth and pour onto macaroni, top with stir-fried pumpkin and minced pork. Serve.

入廚貼士 | Cooking Tips

- 日本南瓜質地軟腍、味甜又色澤金黃，中國南瓜多汁，泰國南瓜則甜美。
- Japanese pumpkin is sweet and sticky with golden brown colour, Chinese pumpkin is juicy while Thailand pumpkin is sweet.

番茄牛肉通粉

Macaroni with Tomato and Beef

材料 | Ingredients

通心粉 100 克	100g macaroni
番茄 4 個	4 tomatoes
牛肉片 150 克	150g sliced beef
上湯 750 毫升	750 ml broth

牛肉醃料 | Beef Marinade

生抽 1 茶匙	1 tsp light soy sauce
生粉 1/2 茶匙	1/2 tsp cornstarch
食用梳打粉 1/8 茶匙	1/8 baking soda
油 1/2 湯匙（後下）	1/2 tbsp oil (add last)
清水 2 湯匙	2 tbsps water

2 人
Serves 2

20 分鐘
20 minutes

⚙️ 調味料 | Seasonings

茄汁 2 湯匙	2 tbsps ketchup
糖 2 茶匙	2 tsps sugar

⚙️ 做法 | Method

1. 通心粉放熱水煮熟，需時約 5~8 分鐘，瀝乾，放入深湯碗中。
2. 在番茄的底部剠「十」字，放入滾水中浸 3 分鐘，撈起，撕皮去核，一半切粒；一半剁茸。
3. 注入上湯、茄汁及糖一同煮滾。
4. 加入番茄茸、牛肉片煮熟，熄火，倒入盛有通心粉的湯碗中，放番茄粒即成。

1. Boil macaroni in hot water for 5-8 minutes. Drain and pour into a deep bowl.
2. Slightly slice crosses on the bottom of tomatoes. Soak in hot water for 3 minutes. Peel skin and remove seeds. Dice half portion and chop another portion into puree.
3. Add broth, ketcup and sugar. Boil until well done.
4. Add tomato puree and sliced beef. Cook until well done. Turn heat off and pour onto macaroni. Sprinkle diced tomato on top and serve.

入廚貼士 | Cooking Tips

- 通心粉可按包裝指示處理，因不同品牌通心粉的軟硬度會有偏差，需要適時檢查熟度。
- The hardnesses of different brands of macaroni are different. Please follow the instruction on packing to cook macaroni and check whether macaroni is well done.

4~6 人
Serve s4~6

20 分鐘
20 minutes

雞絲通粉沙律

Chicken Salad with Macaroni

⬤⬤⬤ 材料 | Ingredients

螺絲粉 225 克
雞肉 225 克
粟米粒 85 克
急凍雜菜粒 55 克
熟雞蛋 2 隻（切粒）
蜜桃 2 個（切粒）
千島沙律汁 225 克

225g macaroni
225g chicken
85g corn
55g frozen mixed vegetables (diced)
2 boiled eggs (diced)
2 pcs peach (diced)
225g Thousand Island salad
dressing

醃料 | Marinade

鹽少許　　　　　　Some salt
胡椒粉少許　　　　Pinch of pepper

做法 | Method

1. 燒一鍋滾水，放入螺絲粉煮約 15 分鐘，不需要蓋蓋，撈起，
   過冷河，待涼。
2. 雞肉洗淨，切粗條，加入醃料拌勻，醃約 30 分鐘。燒熱一鍋水，
   放入雞肉煮熟，撈起，過冷河，待涼。
3. 急凍雜菜粒焯煮片刻，撈起，過冷河，待涼。
4. 所有材料混合，拌入千島沙律汁，置冰箱雪凍即可食用。

1. Boil macaroni in hot water for 15 minutes. Drain, rinse and let
   cool.
2. Wash chicken, cut into strips, marinate for 30 minutes. Boil
   a pot of water, cook chicken in hot water till done, drain and
   rinse. Let cool.
3. Boil frozen mixed vegetables in hot water, drain and rinse. Let
   cool.
4. Mix all ingredients, stir in Thousand Island salad dressing and
   store in fridge. Serve.

入廚貼士 | Cooking Tips
- 所有材料必完全涼凍，方可與千島沙律汁融合，否則沙律很
  容易釋出水份，影響賣相和口感。
- All ingredients must be cool when mixing with Thousand
  Island salad dressing. Otherwise, dressing will give out
  water easily and affect the presentation and taste.

材料 | Ingredients

意大利粉 300 克
豬扒 2 塊
洋葱 1/2 個（切絲）
銀芽 50 克
葱 2 條（切絲）
黑椒碎 2 茶匙
蒜茸 1 茶匙

300g spaghetti
2 pcs pork chop
1/2 onion (shredded)
50g silver sprouts
2 sprigs spring onion (shredded)
2 tsps pepper (diced)
1 tsp minced garlic

2 人
Serves 2

25 分鐘
25 minutes

醃料 | Marinade

生抽 2 茶匙
糖 1 茶匙
鹽 1 茶匙

2 tsps light soy sauce
1 tsp suagr
1 tsp salt

調味料 | Seasonings

生抽 1 1/2 湯匙
老抽 1 湯匙
糖 1 茶匙

1 1/2 tbsps light soy sauce
1 tbsp dark soy sauce
1 tsp sugar

做法 | Method

1. 燒滾水，加入意大利粉及 1 茶匙油，焯煮 8 分鐘，瀝乾待用。

2. 用刀背剁鬆豬扒，下醃料撈勻，醃 2 小時。

3. 燒熱鑊，下油，放入豬扒煎熟，盛起切條備用。

4. 燒熱鑊，下油爆香洋葱絲、蒜茸和黑椒碎，加入意粉、銀芽及調味料，炒勻。

5. 最後加入豬扒條及葱花，快速炒勻即成。

1. Boil spaghetti in hot water with 1 tsp oil for 8 minutes. Drain and set aside.

2. Use back of knife to pound pork chop. Marinate for 2 hours.

3. Heat wok with oil. Pan fry pork chop. Shred and set aside.

4. Heat wok. Saute shredded onion, minced garlic and diced pepper. Add spaghetti, silver sprouts and seasonings and stir-fry.

5. Add shredded pork chop and spring onion. Stir fry-quickly and serve.

入廚貼士 | Cooking Tips

- 熱炒意大利粉前，可用少許熱水或上湯煮焯片刻，比較容易炒勻。

- It is easier to stir-fry spaghetti if cook spaghetti in hot water or broth for a while in advance.

白汁粟米南瓜焗雞球意粉

Baked Spaghetti with Corn, Pumpkin and Chicken in White Sauce

⦿⦿⦿ 材料 │ Ingredients

意大利粉 300 克	300g spaghetti
雞扒 1 塊（約 200 克）	1 pc chicken fillet (200g)
南瓜 10 片	10 slices pumpkin
洋葱粒 2 湯匙	2 tbsps diced onion
粟米粒 2 湯匙	2 tbsps corn
牛油 2 茶匙	2 tsps butter
蒜茸 1 茶匙	1 tsp minced garlic
芝士粉少許	Some cheese powder
鹽少許	Some salt

⦿⦿⦿ 醃料 │ Marinade

生抽 1 茶匙	1 tsp light soy sauce
鹽 1/2 茶匙	1/2 tsp salt
糖 1/2 茶匙	1/2 tsp sugar
黑椒碎 1/2 茶匙	1/2 tsp diced pepper
生粉 1/2 茶匙	1/2 tsp cornstarch
油 1/2 茶匙	1/2 tsp oil

白汁 | White sauce

淡奶 1 杯
麵粉 4 湯匙
牛油 1 湯匙
鹽 1 茶匙
胡椒粉少許
雞粉少許
清水 1 杯

1 cup evaporated milk
4 tbsps flour
1 tbsp butter
1 tsp salt
Pich of pepper
Some chicken powder
1 cup water

入廚貼士 | Cooking Tips

- 用易潔鍋煎雞扒，並輕輕撲上少許麵粉，外皮容易煎至金黃而不黏底。
- Pan fry chicken fillet in non-sticky wok. Dust chicken fillet with some flour to make it easily turns to golden brown colour and do not stick to the wok.

做法 | Method

1. 燒滾水，加入意大粉及 1 茶匙油焯 8 分鐘，瀝乾，加入牛油及少許鹽，放入焗盤待用。
2. 雞扒加入醃料撈勻，並燒熱鑊，下油，放雞扒煎熟，盛起，切件備用。
3. 稍微焯熟南瓜片，備用。
4. 燒熱鑊，下牛油，爆香洋葱粒、蒜茸和粟米粒，再下麵粉，一邊炒一邊分次拌入淡奶、水及其他白汁材料，煮稠成為白汁，熄火。
5. 把南瓜片和雞件鋪在意粉上，淋上白汁，灑上芝士粉，放入已預熱的焗爐，焗至金黃即可。

1. Boil spaghetti in hot water with 1 tsp of oil for 8 minutes. Drain, add in butter and some salt, and put into baking tray.
2. Marinate chicken fillet. Heat work with oil, pan fry chicken fillet until well done. Chop into pieces.
3. Boil sliced pumpkin for a while. Set aside.
4. Heat wok with butter. Saute diced onion, minced garlic and corn. Add flour, stir-fry and add evaporated milk, water and ingredients of white sauce slowly (by batches). Cook until well done (as white sauce). Turn heat off.
5. Put sliced pumpkin and chicken fillet on spaghetti. Top with white sauce and cheese powder. Bake in preheated oven until golden brown.

蠔皇冬菇湯米粉

Rice Vermicelli in Soup with Dried Black Mushroom in Oyster Sauce

⟨⟨⟩⟩ 材料 | Ingredients

乾米粉 150 克
冬菇 6 朵（浸軟，去蒂）
白菜仔 150 克
即食榨菜絲 50 克（略浸洗淨）
上湯適量
薑 2 片

150g dried rice vermicelli
6 dried black mushrooms
 (soaked and stalks removed)
150g white cabbage
50g instant preserved vegetables
 (washed and soaked for a while)
Some broth
2 slices ginger

調味料 | Seasonings

蠔油 1 湯匙	1 tbsp oyster sauce
糖 1 茶匙	1 tsp sugar
麻油 1 茶匙	1 tsp sesame oil
油 1 茶匙	1 tsp oil
清水 100 毫升	100 ml water

芡汁 | Thickening

生粉 1 茶匙	1 tsp cornstarch
清水 3 湯匙	3 tbsps water

做法 | Method

1. 把米粉浸在熱水約 10 分鐘,撈起瀝乾,放入碗中。
2. 燒熱鑊,注入適量熱水。冬菇放碟中,加入適量熱水,以剛蓋過冬菇面為準,加入調味料及薑片,以大火蒸 45 分鐘。
3. 煮沸適量清水,加入少許油和鹽,放入白菜仔焯熟。
4. 熱鑊後燒熱 1 湯匙油,加入即食榨菜絲及 1/2 茶匙糖炒勻,盛起。
5. 將已蒸好的冬菇連汁放鍋中煮沸,倒入芡汁煮稠。
6. 上湯煮沸,倒入已有米粉的碗中,放上榨菜絲、白菜仔和冬菇即可食用。

1. Soak rice vermicelli in hot water for 10 minutes. Drain and pour into a bowl.
2. Heat wok and add hot water. Put dried black mushrooms onto a dish and add hot water to cover dried black mushrooms. Stir in seasoning and sliced ginger and steam for 45 minutes.
3. Boil water with some oil and salt and blanch white cabbage.
4. Heat wok with 1 tbsp of oil. Add instant shredded preserved vegetables and 1/2 tsp sugar, stir-fry and set aside.
5. Boil steamed dried black mushrooms and sauce together. Add thickening and cook until well done.
6. Boil broth, and pour into the bowls with rice vermicelli, place preserved vegetables, white cabbage and dried black mushrooms on top. Serve.

入廚貼士 | Cooking Tips

- 冬菇加入少量生粉及數滴油撈洗,清洗乾淨,可去掉泥味。
- Wash dried black mushrooms with some cornstarch and oil to get rid of mud odor / flavor.

沙嗲牛肉湯米粉

Rice Vermicelli in Soup with Satay Beef

◯◯◯ 材料 | Ingredients

乾米粉 150 克	150g dried rice vermicelli
牛肉 300 克	300g beef
洋蔥 1 個（切絲）	1 onion (shredded)
葱 1 條（切段）	1 sprig spring onion (sectioned)
蒜茸 2 茶匙	2 tsps minced garlic
上湯 750 毫升	750 ml broth

◯◯◯ 調味料 | Seasonings

沙嗲醬 1 1/2 湯匙	1 1/2 tbsps satay paste
生抽 1 茶匙	1 tsp light soy sauce
糖 1/2 茶匙	1/2 tsp sugar

◯◯◯ 芡汁 | Thickening

生粉 1 茶匙	1 tsp cornstarch
清水 3 湯匙	3 tbsps water

⟨⟨⟩⟩ 醃料 | Marinade

生抽 2 茶匙	2 tsps light soy sauce
生粉 1 1/2 茶匙	1 1/2 tsps cornstarch
黑胡椒碎 1 茶匙	1 tsp crushed black pepper
食用梳打粉 1/2 茶匙	1/2 tsp baking soda
油 1 茶匙（後下）	1 tsp oil (add last)
清水 1 湯匙	1 tbsp water

⟨⟨⟩⟩ 做法 | Method

1. 用熱水把米粉浸泡 8 分鐘，盛起，瀝乾，待用。

2. 牛肉洗淨，抹乾，切橫紋薄片，加醃料撈勻，醃 1 小時。

3. 燒熱 1 湯匙油，爆炒洋葱絲，取起。

4. 再燒 2 湯匙油，放下牛肉快炒至八成熟，取起。

5. 用剩餘油，爆香蒜頭和沙嗲醬，加入調味及全部材料，炒勻，勾芡，盛起。

6. 煲滾上湯，加入米粉略煮，放大碗，鋪上沙嗲牛肉即成。

1. Soak rice vermicelli in hot water for 8 minutes. Drain and set aside.

2. Wash beef, pat dry, slice horizontally into thin slices, marinate for 1 hour.

3. Heat wok with 1 tbsp of oil, sauté shredded onion and set aside.

4. Heat wok with 2 tbsps of oil, stir-fry beef until medium well.

5. Sauté garlic and satay paste with the remaining oil, add seasoning and other ingredients. Stir-fry for a while, add thickening and set aside.

6. Bring broth to a boil. Add rice vermicelli and cook for a while. Pour into a bowl and place satay beef on top.

入廚貼士 | Cooking Tips

- 食用梳打粉要與清水調勻，才放入牛肉，效果會更好。
- Dissolve baking soda in water and stir well before adding in beef can have a better result.

4~6 人
Serves 4~6

20 分鐘
20 minutes

星洲炒米

Singapore Style Stir-fried Rice Vermicelli

⦾⦾ 材料 | Ingredients

乾米粉 300 克	300g dried rice vermicelli
叉燒 75 克	75g barbecued pork
（切絲）	(shredded)
銀芽 75 克	75g silver sprouts
蝦仁 40 克	40g shrimps (shelled)
洋蔥 1/4 個（切絲）	1/4 onion (shredded)
三色甜椒各 1/4 個	1/4 red, yellow and green bell
（切絲）	peppers respectively (shredded)
雞蛋 1 隻（打散）	1 egg (whisked)
咖喱粉 1 湯匙	1 tbsp curry powder
魚露 1/2 湯匙	1/2 tbsp fish sauce
黃薑粉 1 茶匙	1 tsp turmeric powder
麻油 1/2 茶匙	1/2 tsp sesame oil

⦾⦾ 蝦醃料 | Marinade for shrimp

蛋白 1/2 茶匙	1/2 tsp egg white
生粉 1/4 茶匙	1/4 tsp cornstarch
鹽 1/8 茶匙	1/8 tsp salt
胡椒粉少許	Pinch of pepper

調味料 | Seasonings

上湯 4 湯匙	4 tbsps broth
鹽 1 茶匙	1 tsp salt
糖 1 茶匙	1 tsp sugar
生抽 1 茶匙	1 tsp light soy sauce
老抽 1 茶匙	1 tsp dark soy sauce
胡椒粉少許	Pinch of pepper

做法 | Method

1. 把米粉放在沸騰的水內浸約 5 分鐘,撈起,過冷河,瀝乾待用。

2. 洗淨蝦仁,抹乾,加醃料醃 10 分鐘。熱鑊燒油,加入蝦仁炒熟備用。

3. 熱鑊燒油,放入 1~2 片薑,倒入銀芽,以大火快炒片刻,讚酒,盛起待用。

4. 雞蛋打散,熱鑊下油煎成蛋皮,待放涼後切絲待用。

5. 熱鑊下油,下洋葱爆香,加入三色椒絲、咖喱粉、黃薑粉、蝦仁和銀芽炒勻,下調味及已浸軟的米粉,拌勻。

6. 待炒乾,放蛋絲炒勻,最後加魚露及麻油拌勻即成。

1. Soak rice vermicelli in hot water for 5 minutes. Drain, rinse in cold water, drain again and set aside.

2. Wash shrimps, pat dry, marinate for 10 minutes. Heat wok with oil, stir-fry shrimps until well done and set aside.

3. Heat wok with oil, ginger slices, add silver sprouts and stir-fry over high heat, sprinkle wine and set aside.

4. Whisk egg. Heat wok with oil and pan fry egg into a sheet. Leave to cool, shred and set aside.

5. Heat wok with oil. Sauté onion, add shredded trio bell peppers, curry powder, turmeric powder, shrimps and silver sprouts and stir-fry for a while. Add seasonings and soaked rice vermicelli, mix well.

6. Stir in shredded egg when the sauce nearly dries up, add fish sauce and sesame oil and serve.

入廚貼士 | Cooking Tips

- 咖喱粉和黃薑粉必須經過熱炒才散發香味。
- Curry powder and turmeric powder should be stir-fried so as to give out aroma.

鹹菜魚鬆燜米

Stewed Rice Vermicelli with Pickled Mustard Green and Shredded Fish

材料 | Ingredients

乾米粉 300 克
鯪魚肉茸（鯪魚膠）150 克
鹹酸菜 40 克
紅辣椒 1 隻（去籽、切絲）
芹菜 1 段（切絲）

300g dried rice vermicelli
150g minced fish
40g pickled mustard green
1 red chili (seeds removed and shredded)
1 section Chinese celery (shredded)

4~6 人
Serves 4~6

15 分鐘
15 minutes

米粉・米線
Rice Vermicelli

⟨⟨⟩⟩ 醬汁 | Sauce

老抽 1 湯匙	1 tbsp dark soy sauce
生抽 1/2 湯匙	1/2 tbsp light soy sauce
糖 1 茶匙	1 tsp sugar
鹽 1/2 茶匙	1/2 tsp salt
胡椒粉少許	Pinch of pepper
麻油少許	Some sesame oil
清水 2 湯匙	2 tbsps water

⟨⟨⟩⟩ 做法 | Method

1. 鹹酸菜浸於清水約 20 分鐘，以鹽擦洗，再用清水沖淨，瀝乾，切碎。

2. 把沸騰的水沖入米粉內浸泡約 5 分鐘，撈起，過冷河，待用。

3. 熱鑊下 2~3 湯匙油，放入鯪魚肉茸煎至兩面金黃，取出切粗條。

4. 再燒熱油，放入芹菜絲及紅椒絲兜炒，加入已浸軟的米粉和鹹酸菜絲略炒，放醬汁炒至收乾，再加入鯪魚條拌勻，即可上碟。

1. Soak pickled mustard green in water for 20 minutes. Rub with salt, rinse, drain and chop.

2. Soak rice vermicelli in hot water for 5 minutes. Drain, rinse in cold water and set aside.

3. Heat wok with 2-3 tbsps of oil. Pan fry minced fish until both sides are golden brown. Take out and cut into strips.

4. Heat wok with oil, sauté shredded Chinese celery and red chili, add soaked rice vermicelli and shredded pickled mustard green, then sauce. Stir-fry until the sauce dries up. Add shredded fish. Mix well and serve.

入廚貼士 | Cooking Tips

- 魚檔售賣的鯪魚肉茸，有些已調味，有些是沒有調味的純魚肉，記得問清楚魚販。
- Remember to ask stall owners whether the minced fish is seasoned or not.

2~4 人
Serves 2~4

10 分鐘
10 minutes

雪菜肉絲燜米粉

Stewed Rice Vermicelli with Potherb
Mustard and Shredded Pork

◯◯◯ 材料 | Ingredients

米粉 400 克
豬肉絲 200 克
銀芽 150 克
雪菜 50 克
冬菇 2 朵（浸軟、切絲）
芹菜 2 條（切絲）
黃椒 1/2 個（切絲）
青椒 1/2 個（切絲）

400g rice vermicelli
200g shredded pork
150g silver sprouts
50g potherb mustard
2 dried black mushrooms (soaked
and shredded)
2 sprigs Chinese celery (shredded)
1/2 yellow bell pepper (shredded)
1/2 green bell pepper (shredded)

⟨⟨⟨ 醃料 | Marinade

鹽 1/2 茶匙
糖 1/2 茶匙
生抽 1/2 茶匙
生粉 1/2 茶匙

1/2 tsp salt
1/2 tsp sugar
1/2 tsp light soy sauce
1/2 tsp cornstarch

⟨⟨⟨ 調味料 | Seasonings

老抽 1 湯匙
生抽 1/2 湯匙
糖 1 1/2 茶匙
清水 1 杯

1 tbsp dark soy sauce
1/2 tbsp light soy sauce
1 1/2 tsps sugar
1 cup water

⟨⟨⟨ 做法 | Method

1. 用清水略浸雪菜，切碎；浸軟米粉。
2. 用醃料放在豬肉絲內拌勻，醃 30 分鐘。
3. 燒熱油，爆香肉絲、冬菇絲、青椒絲、黃椒絲、中國芹菜絲、銀芽和雪菜，備用。
4. 煮滾調味，加入米粉，調慢火煮乾汁液。
5. 最後加入所有材料，炒勻即可進食。

1. Soak potherb mustard in water for a while and chop. Soak rice vermicelli.
2. Marinate shredded pork for 30 minutes.
3. Heat wok with oil, sauté shredded pork, dried black mushrooms, green and yellow bell peppers, Chinese celery, silver sprouts and potherb mustard. Set aside.
4. Bring seasonings to a boil, add rice vermicelli. Adjust to low heat and cook until the sauce dries up.
5. Add all other ingredients, stir-fry and serve.

入廚貼士 | Cooking Tips

- 綠色雪菜的質地爽脆、味道偏淡，黃雪菜的質地略腍、味道偏鹹，需要浸清水久一點。
- Green potherb mustard is crispy with light flavour. Yellow potherb mustard is soft and quite salty which should be soaked in water for longer.

五香雞翼米線

Rice Vermicelli with Five Spices Chicken Wing

◯◯◯ 材料 | Ingredients

雲南米線 180 克
雞翼 445 克（約 12~15 隻）
上湯適量
芫荽適量
葱適量

180g Yunnan rice vermicelli
445g (12-15 pcs) chicken wings
Some broth
Some coriander
Some spring onion

6 人
Serves 6

25 分鐘
25 minutes

醃料 | Marinade

魚露 1 湯匙	1 tbsp fish sauce
生粉 1 湯匙	1 tbsp cornstarch
五香粉 1 茶匙	1 tsp five spices powder
蒜茸 1 茶匙	1 tsp minced garlic
糖 1/2 茶匙	1/2 tsp sugar
麻油少許	Some sesame oil
胡椒粉少許	Pinch of pepper

做法 | Method

1. 燒滾水，把米線煮 5 分鐘，焗燜 15 分鐘，取出過冷河，放進大碗。
2. 把雞翼放醃料內撈勻，醃 2 小時。
3. 燒滾油，放雞翼炸熟，撈起瀝油。
4. 煮滾上湯，注入米線裏，鋪上雞翼，灑上芫荽和葱，即可供食。

1. Boil rice vermicelli in hot water for 5 minutes. Then stew for 15 minutes. Rinse in cold water and pour into a large bowl.
2. Marinate chicken wings for 2 hours.
3. Heat wok with oil, deep-fry chicken wings and drain.
4. Boil broth, pour into rice vermicelli, place chicken wings on top, sprinkle some coriander and spring onion. Serve.

入廚貼士 | Cooking Tips

- 炸雞翼時，必須撈起瀝乾醃汁，否則容易在油炸過程裏濺油，弄傷手臉。
- Drain chicken thoroughly before deep-frying. Otherwise, oil will splash out during deep- frying and may hurt your face and hands.

惹味炸醬米線

Rice Vermicelli with Minced Pork in Bean Sauce

⬭⬭⬭ 材料 | Ingredients

雲南米線 180 克	180g Yunnan rice vermicelli
免治豬肉 240 克	240g minced pork
薑 1 片	1 slice ginger
蒜茸 2 茶匙	2 tsps minced garlic
乾葱茸 2 茶匙	2 tsps shallot puree
豆瓣醬 2 茶匙	2 tsps broad bean sauce
磨豉醬 1 茶匙	1 tsp ground bean sauce
上湯 750 毫升	750 ml broth
芫荽適量	Some coriander
葱粒適量	Some spring onion

⌇ 芡汁 | Thickening

老抽 1 湯匙	1 tbsp dark soy sauce
生抽 1 茶匙	1 tsp light soy sauce
糖 1/4 茶匙	1/4 tsp sugar
鹽 1/8 茶匙	1/8 tsp salt
胡椒粉少許	Pinch of pepper
麻油少許	Some sesame oil
清水 170 毫升	170 ml water

⌇ 做法 | Method

1. 熱鑊下油，爆香薑片、蒜茸和乾葱茸，再放入免治豬肉炒香。
2. 加入磨豉醬及豆瓣醬，炒香後加入芡汁，煮滾，調慢火燜至入味，汁略為收乾。
3. 燒滾水，加入米線煮 5 分鐘，加蓋焗 15 分鐘，取出過冷河，放大碗內。
4. 煮好上湯，注入米線裏，鋪上炸醬，灑上芫荽及葱粒，即可食用。

1. Heat wok with oil, sauté ginger, minced garlic and spring onion. Add minced pork and stir-fry.
2. Add ground bean sauce and broad bean sauce and stir-fry, add thickening and cook for a while. Adjust to low heat and stew until well done (with the sauce nearly dries up).
3. Boil water, add in rice vermicelli and cook for 5 minutes. Cover and boil for 15 minutes. Blanch, drain and pour into a large bowl.
4. Boil broth and pour onto rice vermicelli, place shredded pork with bean sauce on top. Sprinkle coriander and spring onion and serve.

入廚貼士 | Cooking Tips

- 米線需要煮浸焗等過程方會腍軟，否則煮很久麵條仍很硬。
- Soak and boil rice vermicelli to make it soft. Otherwise, it will be hard even after boiling for a long time.

香茅豬扒米線

Rice Vermicelli with Pork Chop and Lemongrass

◯◯◯ 材料 | Ingredients

雲南米線 180 克
豬扒 4 塊
上湯 1 公升
芫茜適量
葱粒適量

180g Yunnan rice vermicelli
4 pcs pork chop
1 litre broth
Some coriander
Some spring onion

4 人
Serves 4

20 分鐘
20 minutes

⊙⊙⊙ 醃料 │ Marinades

香茅 2 枝	2 stalks lemongrass
魚露 1 1/2 湯匙	1 1/2 tbsps fish sauce
生粉 1 湯匙	1 tbsp cornstarch
蒜茸 1 茶匙	1 tsp minced garlic
糖 1/2 茶匙	1/2 tsp sugar
麻油少許	Some sesame oil
胡椒粉少許	Pinch of pepper

⊙⊙⊙ 做法 │ Method

1. 燒滾水，加入米線煮 5 分鐘，加蓋焗 15 分鐘，取出過冷河，放大碗內。
2. 香茅切去頭尾，略爛及切段。
3. 把豬扒放醃料內撈勻，醃 2 小時。
4. 燒滾適量油，放入已醃豬扒，煎熟。
5. 煮滾上湯，倒進米線內，上面鋪上豬扒，灑上芫荽及葱粒，即可供食。

1. Boil water, add rice vermicelli and cook for 5 minutes. Cook with cover for another 15 minutes. Blanch, drain and pour into a large bowl.
2. Cut both ends of lemongrass, pound and chop into sections.
3. Marinate pork chop for 2 hours.
4. Heat wok with oil, add in pork chop and pan fry until done.
5. Boil broth and pour onto rice vermicelli. Place pork chop on top. Sprinkle coriander and spring onion and serve.

入廚貼士 │ Cooking Tips

- 新鮮香茅比香茅粉末的味道清鮮。
- Fresh lemongrass tastes better than lemongrass powder.

4 人
Serves 4

20 分鐘
20 minutes

乾炒牛河

Stir-fried Rice Noodles with Beef

⊙⊙ 材料 | Ingredients

河粉 400 克
牛肉片 150 克
銀芽 150 克
韭黃段 40 克（約 1 吋）
葱 3 條（切絲）

400g rice noodles
150g beef (sliced)
150 silver sprouts
40g chopped chive (about 1 inch)
3 sprigs spring onion (shredded)

⊙⊙ 牛肉醃料 | Beef marinades

生抽 1 茶匙
生粉 1/2 茶匙
食用梳打粉 1/8 茶匙
油 1/2 湯匙（後下）
清水 2 湯匙

1 tsp light soy sauce
1/2 tsp cornstarch
1/8 tsp baking soda
1/2 tbsp oil (add last)
 2 tbsps water

⊙⊙ 調味料 | Seasonings

老抽 4 湯匙
生抽 1 湯匙
糖 1 茶匙

4 tbsps dark soy sauce
1 tbsp oil
1 tsp sugar

⊙⊙ 做法 | Method

1. 將所有醃料用清水調勻，倒入牛肉內拌勻。
2. 燒熱鑊，下油，放入銀芽爆炒片刻，盛起備用。
3. 在鑊中加 2 湯匙油，放入牛肉爆炒片刻，盛起備用。
4. 再加少許油，放入河粉及調味料，炒熱。
5. 放入牛肉和銀芽，炒勻後加入韭黃及葱絲即成。

1. Mix all marinades with water. Pour into beef and mix well.
2. Heat wok with oil. Sauté silver sprout for a while. Set aside.
3. Heat wok with 2 tbsps of oil. Stir-fry sliced beef and set aside.
4. Add some oil. Stir-fry rice noodles and seasonings.
5. Add beef and silver sprouts. Stir-fry, add chive and shredded spring onion and serve.

入廚貼士 | Cooking Tips
- 炒牛河火力控制乃成敗關鍵。
- The control of degree of heat is the key of success of delicious stir-fried rice noodles with beef.

越南炒河

Vietnam Stir-fried Rice Noodles

⊙⊙⊙ 材料 | Ingredients

河粉 300 克
紮肉 115 克
蝦仁 75 克
洋葱絲 75 克
紅椒絲 75 克
紅蘿蔔絲 40 克
乾葱茸 2 茶匙

300g rice noodles
115g Vietnamese sausage roll
75g shrimps (shelled)
75g onion (shredded)
75g red chili (shredded)
40g carrot (shredded)
2 tsps shallot puree

2~4 人
Serves 2~4

10 分鐘
10 minutes

調味料 | Seasonings

魚露 2 茶匙
老抽 1 茶匙
糖 1/4 茶匙
鹽 1/3 茶匙

2 tsps fish sauce
1 tsp dark sauce
1/4 tsp sugar
1/3 tsp salt

做法 | Method

1. 紮肉解凍，沖洗後切粗絲，約 5 毫米 × 5 毫米 × 5 厘米。
2. 熱鑊下 1~2 湯匙油，爆香洋蔥絲、紅椒絲和乾蔥茸，加入蝦仁及紮肉，略炒。
3. 加入河粉、紅蘿蔔絲及調味料，炒勻即可。

1. Defrost Vietnamese sausage roll, wash and shred (about 5 mm × 5 mm × 5 cm).
2. Heat wok with 1-2 tbsps of oil. Sauté shredded onion, red chili and shallot puree. Add shrimps and Vietnamese sausage roll. Stir-fry for a while.
3. Add in rice noodles, shredded carrot and seasonings. Stir-fry well and serve.

入廚貼士 | Cooking Tips

- 買回來的河粉，不要清洗，否則水份太多會容易黏底。
- Do not wash rice noodles. Otherwise, it will contain too much water and stick to bottom of wok easily.

五香牛腩湯河

Rice Noodles in Soup with Five Spices Beef Brisket

◯◯◯ 材料 | Ingredients

河粉 400 克
牛腩 320 克
洋葱 1 個
薑 10 片（汆水和燜用）
紹興酒 2 湯匙
上湯 750 毫升

400g rice noodles
320g beef brisket
1 onion
10 slices ginger (for blanching and stewing)
2 tbsps Shaoxing wine
750 ml broth

調味料 | Seasonings

生抽 2 湯匙	2 tbsps light soy sauce
柱侯醬 1 湯匙	1 tbsp Chu Hou paste
老抽 1 湯匙	1 tbsp dark soy sauce
八角 2 粒	2 pcs anise
片糖約 1/8 片	1/8 pcs brown sugar
清水約 2 杯	2 cups water

做法 | Method

1. 把水煮滾，加入 5 片薑和紹興酒，再放入牛腩煮 15 分鐘，過冷河，切件。
2. 洋葱切件，另加 5 片薑一起放到鍋中爆香。
3. 放進牛腩，爆炒片刻，灒紹興酒，加入調味料煮滾，轉中大火燜 45 分鐘。
4. 將河粉焯煮 10 秒，盛起，瀝乾放碗內，注入上湯，加入五香牛腩即成。

1. Boil water, add in 5 slices of ginger and Shaoxing wine, then beef brisket and stew for 15 minutes. Blanch and chop into pieces.
2. Chop onion, sauté with 5 slices of ginger.
3. Add in beef brisket, stir-fry for a while, sprinkle Shaoxing wine, stir in seasoning and stew for 45 minutes over medium to high heat.
4. Boil rice noodles for 10 seconds and drain. Pour into a bowl, add broth and beef brisket. Serve.

入廚貼士 | Cooking Tips

- 牛腩可用燜燒鍋或真空煲處理，省時兼省能源。
- Electric pressure stew cooker or vacuum pot can be used to stew beef brisket to save time and energy.

橄欖油炒野菌烏冬

Stir-fried Udon with Olive Oil and Mixed Mushrooms

材料 | Ingredients

烏冬 1 包
雞髀菇 50 克
本菇 50 克
秀珍菇 50 克
鮮冬菇 50 克
蒜頭 1 粒
橄欖油 2 茶匙
芝士粉少許

1 pack udon
50g chicken thigh mushrooms
50g clamshell mushrooms
50g oyster mushrooms
50g fresh mushrooms
1 clove garlic
2 tsps olive oil
Pinch of cheese powder

1 人
Serves 1

15 分鐘
15 minutes

醃料 | Marinade

老抽 1 茶匙	1 tsp dark soy sauce
蠔油 1 茶匙	1 tsp oyster sauce
鹽 1/4 茶匙	1/4 tsp salt
糖 1/4 茶匙	1/4 tsp sugar
黑胡椒少許	Pinch of pepper
麻油少許	Some sesame oil

做法 | Method

1. 燒滾水，加入雞髀菇、本菇、秀珍菇和鮮冬菇，稍焯後過冷水，備用。
2. 在鑊中放 2 茶匙橄欖油，先爆香蒜片，隨後加入烏冬，炒熱。
3. 最後加入雞髀菇、本菇、秀珍菇、鮮冬菇和調味料，略炒、上碟後灑上芝士粉即成。

1. Boil chicken thigh mushrooms, clamshell mushrooms, oyster mushrooms and fresh mushrooms in hot water. Blanch and rinse. Set aside.
2. Heat wok with 2 tsps of olive oil. Saute garlic, add in udon, and stir fry.
3. Add in chicken thigh mushroom, clamshell mushroom, oyster mushroom, fresh mushroom and seasonings. Stir fry for a while. Sprinkle cheese powder on top and serve.

入廚貼士 | Cooking Tips
- 把雜菇汆水過冷，可去除本身的異味。
- Blanch mixed mushrooms to remove bad smell / odour.

2 人
Serves 2

15 分鐘
15 minutes

Japanese Curry Beef Udon

日式咖喱牛肉撈烏冬

材料 | Ingredients

肥牛 10 片	10 slices beef
洋葱 1/2 個	1/2 onion
薯仔 1 個	1 potato
甘筍 1/2 條	1/2 carrot
日本咖喱磚 100 克	100g Japanese curry paste
烏冬 2 包	2 packs Udon
上湯 400 毫升	400 ml broth

做法 | Method

1. 燒滾水，加入烏冬煮 2 分鐘，放深碟內。
2. 燒滾水，加入肥牛煮 10 秒，撈起備用。
3. 把薯仔及甘筍切角，炸熟備用。
4. 燒熱鑊，下油，爆香洋葱絲，調慢火，加入咖喱磚爆炒至略溶。
5. 加入已炸熟的薯仔、甘筍及上湯，煮滾後加入肥牛，熄火後拌勻。
6. 把咖喱牛肉鋪在烏冬上，即可供食。

1. Boil Udon in hot water for 2 minutes. Put onto a deep plate.
2. Cook beef in hot water for 10 seconds. Drain and set aside.
3. Cut potato and carrot into wedges. Deep-fry and set aside.
4. Heat wok with oil. Sauté shredded onion. Adjust to low heat. Add in Japanese curry paste. Stir-fry until dissolved.
5. Add in deep-fried potato, carrot, broth and beef. Turn heat off and mix well.
6. Spread curry beef onto Udon and serve.

入廚貼士 | Cooking Tips

- 日式咖喱磚分為甜味、小辣、中辣和勁辣，可按口味選擇。
- There are different flavours of Japanese curry paste including sweet curry, less spicy, medium spicy and very spicy (hot). Choose your own flavour.

乾燒伊麵

Pan fried E-fu Noodles with Soy Sauce

◯◯ 材料 | Ingredients

伊麵 2 個
銀芽 160 克
韭黃 80 克
草菇 10 朵
蒜茸 1 茶匙
薑 2~3 片

2 pcs E-fu noodles
160g silver sprouts
80g yellow chives
10 straw mushrooms
1 tsp minced galirc
2-3 slices ginger

調味料 | Seasonings

蠔油 3 湯匙	3 tbsps oyster sauce
生抽 1 茶匙	1 tsp light soy sauce
糖 1 茶匙	1 tsp sugar
麻油少許	Some sesame oil

做法 | Method

1. 用熱水把伊麵煮軟，瀝乾備用。
2. 洗淨草菇，與薑片放滾水內稍焯，取出，過冷河，開邊。
3. 燒熱油，加入銀芽和韭黃猛火略炒備用。
4. 燒熱油，爆香蒜茸，放下伊麵，慢慢放下拌勻的調味料，並不斷兜炒。
5. 最後加入已處理的銀芽、韭黃和草菇，用大火兜炒 2 分鐘，即可進食。

1. Boil E-fu noodles in hot water until soft. Drain and set aside.
2. Wash straw mushrooms, blanch with ginger in hot water. Take out, blanch and cut into halves.
3. Heat wok with oil. Stir-fry silver sprouts and yellow chives over high heat for a while. Set aside.
4. Heat wok with oil. Sauté minced garlic, add E-fu noodles and seasonings slowly. Stir-fry continuously.
5. Add silver sprouts, yellow chives and straw mushrooms at last. Stir-fry over high heat for 2 minutes. Serve.

入廚貴士 | Cooking Tips

- 伊麵經油炸，比較油膩，需要焯煮過冷，把油份去掉。
- Deep-fried E-fu noodles are pretty oily. Blanch and rinse to remove oil before cooking.

吉列豬扒拉麵

Ramen with Cutlet Pork Chop

拉麵 1 包
豬扒 2 塊
蛋黃 2 隻
生粉適量
麵包糠適量
上湯 750 毫升
油 2 杯

1 pack ramen
2 slices pork chop
2 egg yolks
Some cornstarch
Some breadcrumbs
750 ml broth
2 cups oil

1~2 人
Serves 1~2

10 分鐘
10 minutes

78

⑩⑩ 醃料 | Marinades

雞蛋白 1 隻	1 egg white
黑胡椒粉 2 茶匙	2 tsps pepper
生抽 1 茶匙	1 tsp light soy sauce
糖 1 茶匙	1 tsp sugar
鹽 1/2 茶匙	1/2 tsp salt

⑩⑩ 做法 | Method

1. 用刀背剁鬆豬扒，放入醃料撈勻醃 2 小時。
2. 在豬扒上撲生粉，再沾滿蛋黃液，然後沾滿麵包糠，輕輕壓緊。
3. 燒熱油，把豬扒炸至金黃，撈起瀝油。
4. 煮滾上湯，放入拉麵，煮熟後倒在碗中，放上吉列豬扒，即可食用。

1. Pound pork chop with the back of knife, then marinate for 2 hours.
2. Dust some flour onto pork chop, dip some egg yolks and breadcrumbs and slightly press.
3. Heat wok with oil, deep-fry pork chop until golden brown and drain.
4. Boil broth, add ramen and pour into a bowl after well done. Put cutlet pork chop on top and serve.

入廚貼士 | Cooking Tips

- 豬扒先撲粉封鎖肉面，保存肉汁。選用粗粒麵包糠可增強嚼勁。
- Dust some flour onto pork chop to retain the gravy. Coarse breadcrumbs can be used to increase chewing texture.

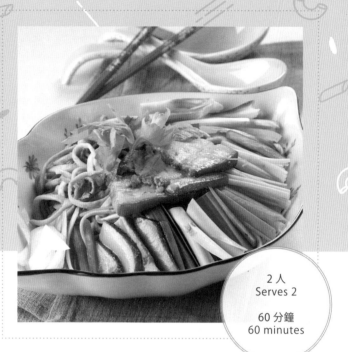

2 人
Serves 2

60 分鐘
60 minutes

天長地久炒麵
Long Lasting Stir-fried Noodles

⟨⟨⟩⟩ 材料 | Ingredients

伊麵 2 個	2 pcs E-fu noodles
五花腩 300 克	300g pork belly
蝦仁 80 克	80g shrimps (shelled)
葱 2 條（切絲）	2 sprigs spring onion (shredded)

⟨⟨⟩⟩ 雜菜(切絲) | Mixed vegetables (shredded)

冬菇 2 朵	2 dried black mushrooms
洋葱 1/2 個	1/2 onion
青椒 1/2 個	1/2 green pepper
紅椒 1/2 個	1/2 red pepper
芹菜 2 條	2 stalks Chinese celery
甘筍 5 片	5 slices carrot

⟨⟨⟩⟩ 調味料 | Seasonings

蠔油 3 湯匙	3 tbsps oyster sauce
生抽 1 茶匙	1 tsp light soy sauce
糖 1 茶匙	1 tsp sugar
麻油少許	Some sesame oil

⊙⊙ 燜豬肉料 | Ingredients for stewing pork belly

薑 1 片	1 slice ginger
生抽 3 湯匙	3 tbsps light soy sauce
老抽 1 湯匙	1 tsp dark soy sauce
蒜茸 1 茶匙	1 tsp minced garlic
磨豉醬 1 茶匙	1 tsp ground bean sauce
冰糖 1 茶匙（壓碎）	1 tsp rock sugar (crushed)
八角 1 粒	1 anise
清水 3/4 杯	3/4 cup water

⊙⊙ 做法 | Method

1. 伊麵放熱水內煮軟，過冷河，瀝乾。
2. 把原塊五花腩汆水。
3. 燒熱油，爆香薑片和蒜茸，加入磨豉醬炒香，再加入其他燜豬肉料煮滾。調慢火，燜45分鐘，取出，放涼，切薄片。
4. 燒熱油，爆香雜菜絲，再放下蝦仁炒熟，加入伊麵、葱絲及拌勻的調味料，用大火兜炒 2 分鐘，熄火盛起。
5. 將五花腩肉片鋪在炒好的伊麵上面，即可食用。

1. Boil E-fu noodles in hot water until soft. Blanch and drain.
2. Blanch pork belly and set aside.
3. Heat wok with oil, sauté ginger and minced garlic. Add ground bean sauce and stir-fry, add other ingredients for stewing pork belly and cook for a while. Adjust to low heat and stew for 45 minutes. Take out, let cool and slice into thin slices.
4. Heat wok with oil, sauté shredded mixed vegetables. Add shrimps, E-fu noodles, shredded spring onion and seasonings. Stir-fry over high heat for 2 minutes. Turn heat off and set aside.
5. Place sliced pork belly onto stir-fried E-fu noodles. Serve.

入廚貼士 | Cooking Tips

- 剛燜好的豬腩肉不會很腍軟，熄火後原鍋焗 15 至 20 分鐘會比較好。
- It is better to keep the stewed pork belly for 15-20 minutes with cover after heat is turned off so that it could be sticky and soft.

2~4 人
Serves 2~4

20 分鐘
20 minutes

三絲湯年糕

Rice Cake in Soup with Shredded Pork, Ham and Preserved Vegetables

⟨⟨⟩⟩ 材料 | Ingredients

上海年糕 4 條
肉絲 200 克
火腿絲 80 克
榨菜絲 50 克
雞蛋 2 隻
上湯適量

4 pcs Shanghai rice cake
200g shredded pork
80g shredded ham
50g preserved vegetables
2 eggs
Some broth

⊗⊗ 調味料 | Seasonings

生抽 2 茶匙   2 tsps light soy sauce
生粉 2 茶匙   2 tsps cornstarch
胡椒粉少許   Pinch of pepper
麻油少許   Some sesame oil

⊗⊗ 做法 | Method

1. 上海年糕放冷水中浸一夜，取出切片，再用熱水煮熟，撈起，放進湯碗。

2. 洗淨榨菜絲，略浸。肉絲與調味料撈勻醃 30 分鐘。

3. 打勻蛋漿，在鑊中燒熱油，倒入蛋漿慢慢煎成蛋皮，盛起待涼後切絲，待用。

4. 燒熱油，炒熟肉絲，再加入榨菜、火腿絲及 1/2 茶匙糖，兜炒，盛起備用。

5. 煮滾上湯，倒入盛有上海年糕的湯碗中，加入已炒好的三絲，再鋪上蛋皮絲，即可食用。

1. Soak Shanghai rice cake in cold water overnight. Take out and slice, boil in hot water, drain and pour into a bowl.

2. Wash preserved vegetables and soak for a while. Marinate with seasonings for 30 minutes.

3. Whisk egg. Heat wok with oil, add whisked egg and pan fry. Leave to cool and shred. Set aside.

4. Heat wok with oil, stir-fry shredded pork, add preserved vegetables, shredded ham and 1/2 tsp of sugar. Stir-fry until well done. Set aside.

5. Boil broth and pour into a bowl with rice cake. Add stir-fired shredded pork, ham and preserved vegetables. Sprinkle shredded eggs on top. Serve.

入廚貼士 | Cooking Tips
* 煎蛋皮時，鑊要夠熱，油要多一點，才容易煎出蛋皮。
* Pan fry egg in a heated wok with much oil.

◯◯◯ 材料 | Ingredients

上海年糕 4 條	4 pcs Shanghai rice cake
豬肉絲 150 克	150g shredded pork
白菜仔 150 克	150g cabbage
冬筍 40 克	40g winter bamboo shoots
甘筍片 40 克	40g sliced carrot
薑 1 片	1 slice ginger

◯◯◯ 豬肉絲醃料 | Marinade for Pork

油 1/2 湯匙	1/2 tbsp oil
鹽 1/4 茶匙	1/4 tsp salt
糖 1/4 茶匙	1/4 tsp sugar
生粉 1/3 茶匙	1/3 tsp cornstarch
胡椒粉少許	Pinch of pepper
麻油少許	Some sesame oil
清水 1 湯匙	1 tbsp water

3~4 人
Serves 3~4

10 分鐘
10 minutes

◯◯ 芡汁 | Thickening

上湯 75 毫升	75 ml broth
蠔油 1 茶匙	1 tsp oyster sauce
鹽 1/4 茶匙	1/4 tsp salt
糖 1/4 茶匙	1/4 tsp sugar
胡椒粉少許	Pinch of pepper
麻油少許	Some sesame oil

◯◯ 做法 | Method

1. 浸軟上海年糕，切片，稍焯，待用。
2. 豬肉絲放入醃料內撈勻，醃 15 分鐘，稍焯，待用。
3. 冬筍片和甘筍片稍焯待用。
4. 洗淨白菜仔，稍焯，待用。
5. 燒熱鑊，下油，爆香薑片，加入所有材料，略炒，加入芡汁煮 5 分鐘即成。

1. Soak Shanghai rice cake, slice, boil for a while and set aside.
2. Marinate shredded pork for 15 minutes, blanch for a while and set aside.
3. Boil winter bamboo shoots and carrot slices for a while. Set aside.
4. Wash cabbage, blanch for a while and set aside.
5. Heat wok with oil. Sauté ginger slices, add all ingredients and stir-fry, add thickening and cook for 5 minutes.

入廚貼士 | Cooking Tips

- 年糕容易黏底，所以用汁烹煮效果比較好。
- As rice cake is easily sticks to the bottom of wok, it is better to cook it with sauce.

四川撈粉皮

Sichuan Sheet Jelly

材料 | Ingredients

乾綠生粉皮 150 克
脆炸花生 3 湯匙（壓碎）
蒜茸 2 湯匙
葱 1 棵（切粒）
辣椒 2~3 隻（切粒）

150g dried green raw sheet jelly
3 tbsps deep-fried peanuts (grinded)
2 tbsps minced garlic
1 sprig spring onion (diced)
2-3 pcs red chili (diced)

◎◎◎ 四川辣椒汁 │ Sichuan chili sauce

蒜茸 100 克	100g minced garlic
川椒粒（即花椒粒）20 克	20g Sichuan peppercorn
辣椒乾 50 克（切粒）	50g dried red chili (diced)
油 250 毫升	250 ml oil

◎◎◎ 調味料 │ Seasonings

鹽 1 1/2 茶匙	1 1/2 tsps salt
雞粉 1 茶匙	1 tsp chicken powder

◎◎◎ 做法 │ Method

1. 油燒至八成熱，放入蒜茸炸至金黃色，熄火，放入川椒粒和辣椒乾粒浸至出味。
2. 用凍水浸泡乾綠生粉皮約 15~20 分鐘，讓其變軟，放滾水浸片刻，撈起。
3. 把粉皮和適量四川辣椒汁拌勻，撒上花生碎、葱粒和辣椒粒，即可。

1. Heat wok with oil until 80% heat. Deep-fry minced garlic until golden brown. Turn heat off. Add Sichuan peppercorn and diced red chili and soak for a while until aroma / flavor comes out.
2. Soak dried green raw sheet jelly in cold water for 15-20 minutes until soft. Soak in hot water for a while. Drain.
3. Mix sheet jelly with Sichuan chili sauce. Sprinkle grinded peanuts, diced spring onion and red chili on top. Serve.

入廚貼士 │ Cooking Tips

- 四川辣椒汁煮成後，放入密封玻璃瓶內置陰涼地方貯藏。
- Sichuan chili sauce could be stored into an airtight container and keep in cool and dim place.

簡易粉麵

編著
徐嘉儀

編輯
紫彤

美術設計
Nora Chung

排版
萬里機構製作部

翻譯
Sa

攝影
Lasso Adv. Agency

出版者
萬里機構出版有限公司
香港鰂魚涌英皇道1065號東達中心1305室
電話：2564 7511
傳真：2565 5539
電郵：info@wanlibk.com
網址：http://www.wanlibk.com
　　　http://www.facebook.com/wanlibk

發行者
香港聯合書刊物流有限公司
香港新界大埔汀麗路36號
中華商務印刷大廈3字樓
電話：2150 2100
傳真：2407 3062
電郵：info@suplogistics.com.hk

承印者
美雅印刷製本有限公司

出版日期
二零一八年八月第一次印刷

萬里機構

萬里 Facebook